I0475285

The Practically Cheating Statistics Handbook

TI-89 Companion Guide

S. Deviant, MAT

StatisticsHowTo.com

info@statisticshowto.com

Printed in the United States of America

ISBN 1-45-379811-0

Contents

Basics

1. Mean, Median, and Mode

Sample problem: Find the mean, mode, and median for the following list of numbers: 1, 9, 2, 3, 7, 8, 9, 2.

Step 1: Press `APPS` then scroll to **Stats/List Editor** (scroll with the arrow keys at the top right of the keypad). Press `ENTER`.

Step 2: *Clear any data in the list editor by pressing* `F1` *then* `8`.

Step 3: *Press* `ALPHA` , `5` *, and* `ENTER`. This names your list "m." Make sure m appears in the field: if 5 appears, it means the `ALPHA` key didn't work: try it again.

Step 4: *Enter your numbers, one at a time. Follow each entry by pressing the* `ENTER` *key.* For our group of numbers, enter:

`1` `ENTER`
`9` `ENTER`
`2` `ENTER`
`3` `ENTER`
`7` `ENTER`
`8` `ENTER`
`9` `ENTER`
`2` `ENTER`

Step 5: *Press* `F4` *, then* `ENTER` (for the 1-var stats screen).

Step 6: *Tell the calculator you want stats for the list called "m" by* ENTER *ing* ALPHA , 5 *into the "List:" box.* The calculator should automatically put the cursor there for you. Press ENTER ENTER .

Step 7: *Read the results for the mean.* The mean is the first in the list (an x with a bar on top), = **5.125**.

Step 8: *Read the results for the median:* The median is about half way down the list: scroll with ↓ and look for MedX = **5**.

Step 9: *Find the mode:* Press ENTER to return to the list editor. Press F3 2 , ENTER . Make sure "m" is in the "List:" box and the order is "Ascending." Press ENTER . Your data is now sorted. Just count which number appears the most: that's your mode.

Tip #1: Press ↓ to scroll down the complete list: only a partial list will appear on screen in the List Editor.

Tip #2: You can name your list anything you want, but keep it simple and don't use common variables like t, x, y, or z.

Tip #3: To change the list order to ascending from descending, use ↓ to scroll down and → to bring up the menu. Use the arrow keys to choose, then press ENTER .

2. Interquartile Range

Sample problem: Find Q1, Q3, and the interquartile range for the following list of numbers: 1, 9, 2, 3, 7, 8, 9, 2.

Step 1: *Press* `APPS`. *Scroll to Stats/List Editor (use the arrow keys on the keypad to scroll). Press* `ENTER`.

Step 2: *Clear the list editor of data: press* `F1` `8`.

Step 3: *Press* `ALPHA`, `9`, `ALPHA`, `1`, *and* `ENTER`. *This names your list "IQ."*

Step 4: *Enter your numbers, one at a time. Follow each entry by pressing the* `ENTER` *key. For our group of numbers,* `ENTER`

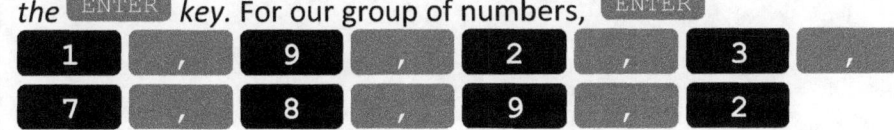

Step 5: *Press* `F4`, *then* `ENTER` *(for the 1-var stats screen).*

Step 6: *Tell the calculator you want stats for the list called "IQ" by* `ENTER`*ing* `ALPHA`, `9`, `ALPHA`, `1` *into the "List:" box. The calculator should automatically put the cursor there for you.* Press `ENTER`.

Step 7: *Read the results.* Q1 is listed as QlX (in our example, QlX = 2). Q3 is listed as Q3X (Q3X = 8.5). To find the interquartile range, subtract Q3 from Q1. The interquartile range is

8.5–2 = 6.5

3. Variance

Sample problem: Find the Variance for the following data set: 2, 4, 5, 6, 8, 10, 15, 23.

Step 1: Press the HOME key. The HOME key is on the left hand side, third button from the top.

Step 2: Press the CATALOG key.
It's located below the APPS key in the top middle of your keypad.

Step 3: Press ↓ to scroll to **variance(**. Press ENTER .

Step 4: Press 2ND (. If you did this correctly, you should have a curly bracket. The following should now be on your screen:
variance({

Step 5: Enter your data set.
In our example, ENTER

2	,
4	,
5	,
6	,
8	,
1	0	,
1	5	,
2	3

So you get: **variance({2,4,5,6,8,10,15,23**

Step 6: Press, `2ND` `)` `)` These keystrokes close the expression so you have:

variance({2,4,5,6,8,10,15,23}). (**Important!** Your expression must look exactly like this, with both sets of parentheses and curly parentheses.)

Step 7: Press `ENTER`.

Step 9: Press `♦` and then `ENTER` to get a decimal answer if the calculator gives you a radical or fractional expression (and it probably will).

That's it! The variance for the data set {2, 4, 5, 6, 8, 10, 15, 23} is displayed, which is **47.5536**.

4. Standard Deviation

Sample problem: Find the Standard Deviation for the following data set:
1, 34, 56, 89, 287, 598, 1001

Step 1: *Press the* HOME *key. The* HOME *key is on the left hand side, third button from the top.*

Step 2: *Press the* CATALOG *key. It's located below the* APPS *key in the top middle of your keypad.*

Step 3: Scroll down to **stdDev(** using ⬇. Press ENTER.

Step 4: *Press* 2ND , (.The following should be on screen (note the curly bracket you just added): stdDev({

Step 5:*Enter your set of numbers*. In our example, ENTER

1 ,

3 4 ,

5 6 ,

8 9 ,

2 8 7 ,

5 9 8 ,

1 0 0 1

You should have the following displayed on screen:
stdDev({1,34,56,89,287,598,1001

Step 6: *Press* [2ND] [ )] [ )]. This adds a parentheses and curly parentheses to close the expression so you have **stdDev({1,34,56,89,287,598,1001})**. Important! Your expression must look exactly like this, with both sets of parentheses and curly parentheses.

Step 7: *Press* [ENTER]. The standard deviation displayed is 375.149.

Tip: If you have a radical or fractional expression on screen (2√1724030/7) and require a decimal, just press [♦] and [ENTER] to get a decimal.

Graphs

1. Box Plot and Five Number Summary

When you create a box plot with whiskers on the TI-89, the TI-89 will automatically calculate the five number summary for you, saving you time and effort.

Sample problem: Create a boxplot and find the five number summary for the following data: 200, 350, 300, 350, and 400.

Step 1: Create a new folder called "Box." From the HOME screen, press F4 and scroll down to **NewFold** (option B). Press ENTER.

Step 2: Enter 2ND ALPHA (– x to spell B O X and press ENTER.

Step 3: Press APPS then scroll down to **Stats/List Editor**. Press ENTER.

Step 4: Press ↓ to get to the first line of the list. Enter your data into list1. Follow each entry with a comma: 200, 350, 300, 350, 400.

Step 5: Press F2, 1 to ENTER **Plot Setup**.

Step 6: Press F1, right arrow, and 5 to select **mod box plot**.

Step 7: Arrow down to **Mark** and select **box**.

Step 8: Arrow down and  ENTER B O X in the **x** box 2ND ALPHA (− x to spell B O X. Press ENTER .

Step 9: Read the boxplot. Press F3 and use the left and right cursors to find Min(200), Q1(250), Med(325), Q3(400), and Max(500).

Tip: if you want to change the folder back to MAIN, press MODE , scroll down to **Current Folder**. Press right key, then press 1 ENTER .

2. Cumulative Frequency Table

Sample question: Build a cumulative frequency table for the following classes.

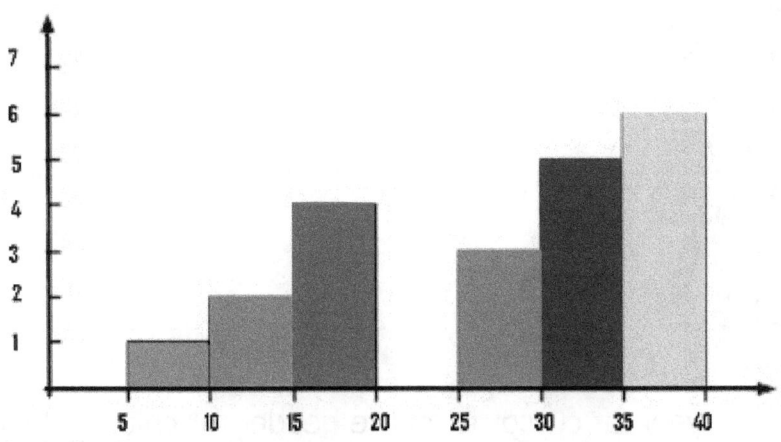

Class Limits	Frequency
5-10	1
10-15	2
15-20	4
20-25	0
25-30	3
30-35	5
35-40	6

Step 1: Press APPS and scroll to **Stats/List Editor**. Press ENTER.

Step 2: Press F1 8 to clear any data in the editor.

Step 3: Name your first column "L1" by entering ALPHA 4 1.

Step 4: Enter your values into L1, following each number by an Enter key:

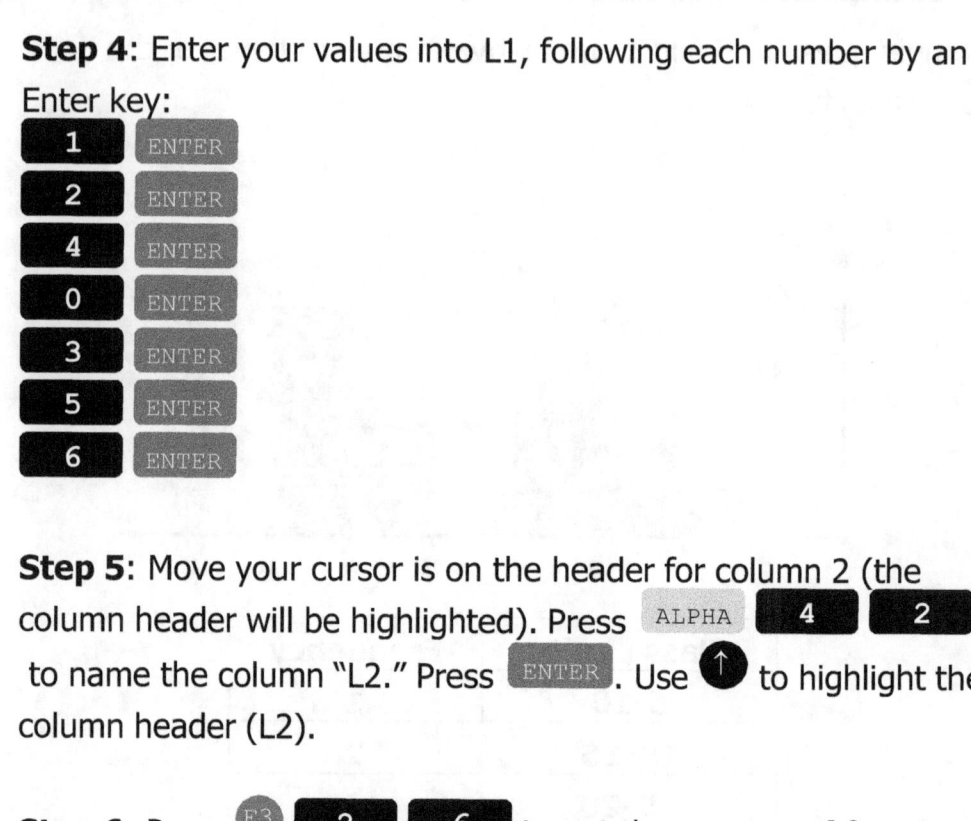

Step 5: Move your cursor is on the header for column 2 (the column header will be highlighted). Press ALPHA 4 2 to name the column "L2." Press ENTER. Use ↑ to highlight the column header (L2).

Step 6: Press F3 2 6 to get the **cumsum(** function.

Step 7: Enter "L1" into the cumsum function by pressing ALPHA 4 1. Press the) key then press ENTER.

Step 8: The list of cumulative frequencies for each value in L1 are returned in L2: **1, 3, 7, 7, 10, 15**.

Tip: it doesn't matter what the columns are called or headed. Just make sure you paste the header for column into the **cumsum(** function.

---

3. Frequency Chart or Histogram

Sample problem: Create a frequency distribution chart for the following new car costs: 12,500; 22,400; 14,300; 32,200; 21,500; 19,980; 15,001; 22,001; 32,036; 35,124; 29,001; 25,006; 27,001; and 18,500.

Step 1: Press APPS and scroll to the Stats/List Editor. Press ENTER.

Step 2: Press F1 then 8 to clear the list editor of data.

Step 3: Enter "cars" as the name by pressing 2ND ALPHA) = 2 3 ALPHA ENTER.

Step 4: Enter your data:

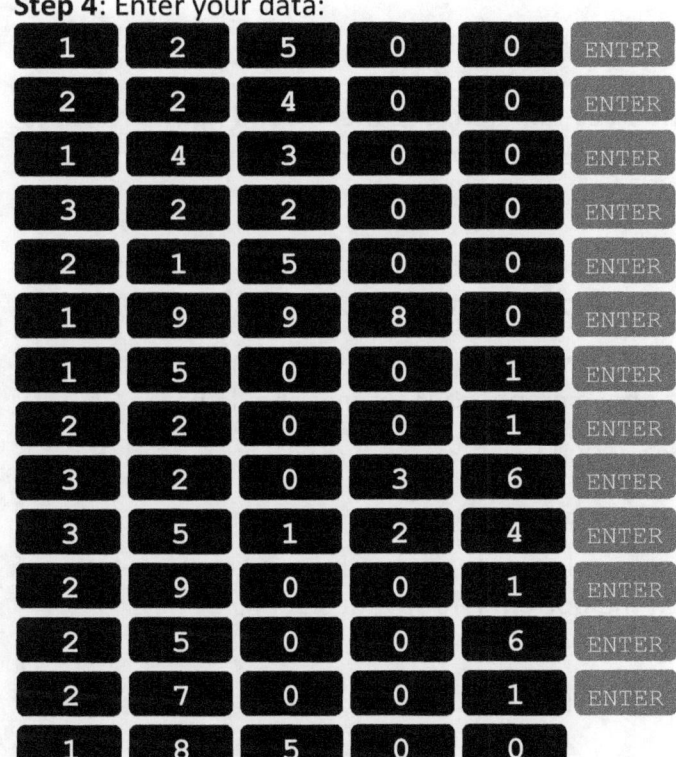

1	2	5	0	0	ENTER
2	2	4	0	0	ENTER
1	4	3	0	0	ENTER
3	2	2	0	0	ENTER
2	1	5	0	0	ENTER
1	9	9	8	0	ENTER
1	5	0	0	1	ENTER
2	2	0	0	1	ENTER
3	2	0	3	6	ENTER
3	5	1	2	4	ENTER
2	9	0	0	1	ENTER
2	5	0	0	6	ENTER
2	7	0	0	1	ENTER
1	8	5	0	0	

Step 5: Press F2 ENTER and F1 to go into Plot Setup (Define Plot).

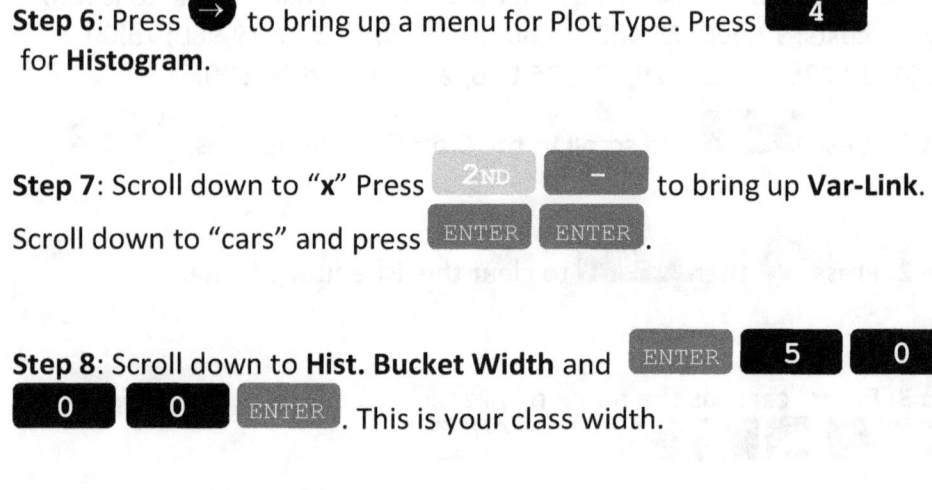

Step 6: Press ➡ to bring up a menu for Plot Type. Press **4** for **Histogram**.

Step 7: Scroll down to "**x**" Press **2ND** **–** to bring up **Var-Link**. Scroll down to "cars" and press **ENTER** **ENTER**.

Step 8: Scroll down to **Hist. Bucket Width** and **ENTER** **5** **0** **0** **0** **ENTER**. This is your class width.

Step 9: Press **ENTER** **F5**.

Step 10: Press **F3** for the trace function. Use the left and right scroll keys to move from one bar to another. (This will tell you how many items are in each class (n = x)).

Tip #1: If your graph doesn't show (or only part of the graph shows), you may need to change the window. Press **♦** **F2** to check your window settings. For the above graph, your settings should be approximately 10000 < x < 40000 and 0 < y < 8. Press **F2** **9** to return to the graph.

Tip #2: Make sure the **ALPHA** lock is turned on by checking for a little black box with "a" in it on the bottom left of your screen.

4. Scatter Plot

Sample problem: Create a scatter plot for the following coordinates (1, 3), (2, 4), (3, 9), (4, 11), and (5, 12).

Step 1: Go into the **Stats/List Editor**. To get there, press `APPS`, scroll down to the **Stats/List Editor**, and press `ENTER`. *Note*: If you are asked for a folder, the default is main. Just press `ENTER`.

Step 2: Clear the List Editor. Press `F1` and `8`.

Step 3: Enter "SP" as the variable name by pressing `ALPHA` `3` `ALPHA` `STO▶`.

Step 4: *Enter your x-variables, one at a time.* Follow each number by pressing the `ENTER` key. For our list, you would `ENTER`:

1	ENTER
2	ENTER
3	ENTER
4	ENTER
5	ENTER

Step 5: Use the arrow keys to scroll across to the next column's header (if you get stuck in a weird menu, press `ESC`. Enter the variable name SP2 by pressing `ALPHA` `3` `ALPHA` `STO▶` `2`.

Step 6: Enter your y-variables, one at a time. Follow each number by pressing the ENTER key. For our list, you would enter:

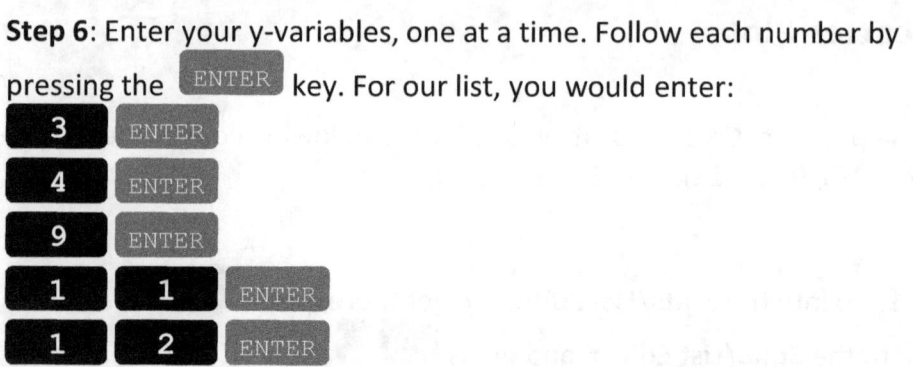

Step 7: Press F2 ENTER F1 . Make sure "Scatter" shows at the top. If it does not, use → to change it, then press ENTER . Scroll down. Enter "SP" in the **X** box and "SP2" in the **Y** box (using the same keystrokes as Steps 3 and 5. Press ENTER ENTER ENTER ♦ F1 F2 9 .

Probability and Binomial Distributions

1. Combinations

Sample problem: If there are 5 people, Barb, Sue, Jan, Jim, and Rob, and only three will be chosen for the new Parent Teacher Association, how many combinations are possible for the committee?

Step 1: Press the HOME screen on the calculator.

Step 2: Press the CATALOG key.

Step 3: Press ALPHA 6 . This selects the letter "n" and brings you to the "n"s in the list. If it doesn't, try pressing ALPHA 6 again.

Step 4: Find **nCr(**. Press ENTER .

Step 5: Press 5 . This is the number of possible people, or n.

Step 6: Press , then 3 . This is the number of people we need to choose, or r.

Step 7: Press) . The entry on your calculator should now read **nCr(5,3)**. Press ENTER . This returns you result. There are **10** possible ways this committee can be chosen.

Tip: Instead of hitting the ALPHA and 6 keys, use the scroll button to scroll through the menu.

---

2. Binomial Probability

Sample problem: John's batting average is .240. If he's at bat three times, what is the probability that he gets exactly three hits?

Step 1: Press `APPS` and scroll (using the scrolling arrows) to choose **Stats/List Editor**. Press `ENTER`.

Step 2: Press `F5`. Scroll down to **B: Binomial Pdf**. Press `ENTER`.

Step 3: Enter the number of trials. John bats three times, so the number of trials is **3**. Press `3` and hit ↓.

Step 4: Enter the Probability of Success, **P**. John's batting average is **.240**, so enter .240. Press ↓.

Step 5: Enter the X value. We want to know the probability of John getting exactly three hits, so enter `3` in the X Value box.

Step 6: Press `ENTER` for the result. The probability of John getting exactly three hits is **.150072**, returned at the top of the screen as "**Pdf=.150072**".

Tip: Instead of scrolling down to Binomial Pdf, you can hit the `ALPHA` and `(` keys to select it instead.

Warning: BinomialPdf is an exact probability for one value of x. If you want to find a cumulative probability (for example, what are John's chances of getting 0 or 1 hits?) you will need the use the BinomialCdf function.

3. Binomial Probability: BinomialCdf

Sample problem: Jane's batting average is .230. If she's at bat four times, what is the probability that she gets three or four hits?

Step 1: Press APPS and use the scroll arrows to highlight the **Stats/List Editor**. Press ENTER.

Step 2: Press F5. Press ALPHA and). This should bring up the **Binomial Cdf** screen. If it doesn't, make sure you pressed down the ALPHA key (you are using it to choose "C" above the) key).

Step 3: Enter the number of trials in the **Num Trials** box. Jane goes to bat four times, so the number of trials is 4. Press 4 and then ↓.

Step 4: Enter the probability: "Prob Success, p." Jane's batting average is .230, so ENTER . 2 3 0 in this box. Press ↓.

Step 5: Enter the lower value, 3. You want to know the probability of Jane getting between three and four hits, so you ENTER 3 in the X Value box.

Step 6: Enter the upper value, 4.

Step 7: Press ENTER for the result. The probability of Jane getting three or four hits is **.040273**, returned at the top of the screen as **"Pdf=.040273"**.

Tip: Instead hitting the ALPHA and   keys to select BinomiaiCdf, you can scroll down the menu with the arrow keys instead.

Warning: Use the BinomiaiPdffunction (option B from the **fl** menu) if you only have one x-value.

4. Mean and Standard Deviation for a Binomial Distribution

Sample problem: Find the mean and standard deviation for a binomial distribution with n = 5 and p = 0.12.

Step 1: Press `APPS` and select the **Stats/List Editor**. Press `ENTER`.

Step 2: Press `F1` and `8` to clear the list editor.

Step 3: Name the first column "Bin" by entering `2ND` `ALPHA` (for `ALPHA` lock) then `(` `9` `6`. Press `ENTER`.

Step 4: Enter the following into column 1:

`0` `ENTER`
`1` `ENTER`
`2` `ENTER`
`3` `ENTER`
`4` `ENTER`
`5` `ENTER`

Step 5. Press `F5` and scroll down to **BinomialPdf (option B)**. Press `ENTER`.

Step 6: Enter `5` into the **Num Trials, n** box.

Step 7: Scroll down and  into the **Prob Success, p** box.

Step 8: Scroll down and delete anything in the **X** box to leave a blank value. Press `ENTER`.

Step 9: Press `ENTER` again. The BinomialPdf values are entered into a list title "PDF."

Step 10: Press `F4` `ENTER`. Type "bin" into the "List" box using the same keystrokes from step 3.

Step 11: Type "statvars\pdf" into the **Freq** box (press 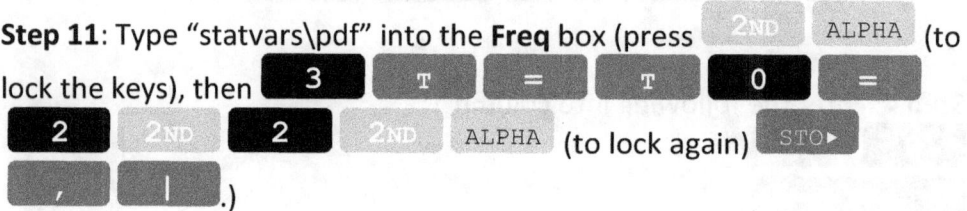.)

Step 12: Press `ENTER` and read the result. The mean (the top value, an x with a bar on top) is **.06**. The standard deviation is **.243475** and is the fifth line down (σx).

Tip: If you get an error like "argument mismatch," check your inputs in the 1-var stats box to make sure they say "bin" and "statvars\pdf" without the quotes.

Normal Distributions

1. Central Limit Theorem

In statistics mumbo-jumbo, they say that if **n × p** and **n × (1 − p)** are greater than 10, then you can use the normal approximation to the binomial. What this means to you is that if you have a large sample size, or if the question hints at a normal approximation, then you can use this technique on the TI-89.

Sample problem: A population of community college students includes inner city students (p = .33). What is the probability that a random sample of 45 students from the population will have from 20% to 40% inner city students?

Step 1: Press APPS . Highlight the **Stats/List Editor** by using the scroll keys. Press ENTER .

Step 2: Press F5 and scroll down to C: **BinomialCdf**.

Step 3: Enter 4 5 in the **Num Trials** box.

Step 4: Scroll down and 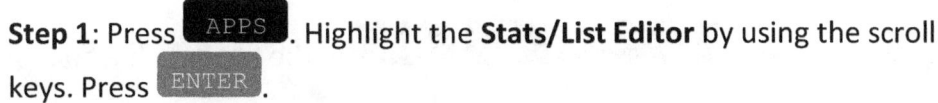 in the **Prob Success** box.

Step 5: Scroll down and 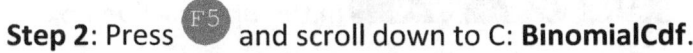 in the **Lower Value** box (because 20% of 45 = 9).

Step 6: Scroll down and  in the **Upper Value** box (because 40% of 45 = 18). Press ENTER .

Step 7: Read the Result: **Cdf=.857142**. This means that the probability your random sample will have 20-40% inner city students is **85.71%**.

2. Normally Distributed Probability Question

Sample question: A group of students with normally distributed salaries earn an average of $6,800 with a standard distribution of $2,500. What proportion of students earn between $6,500 and $7,300?

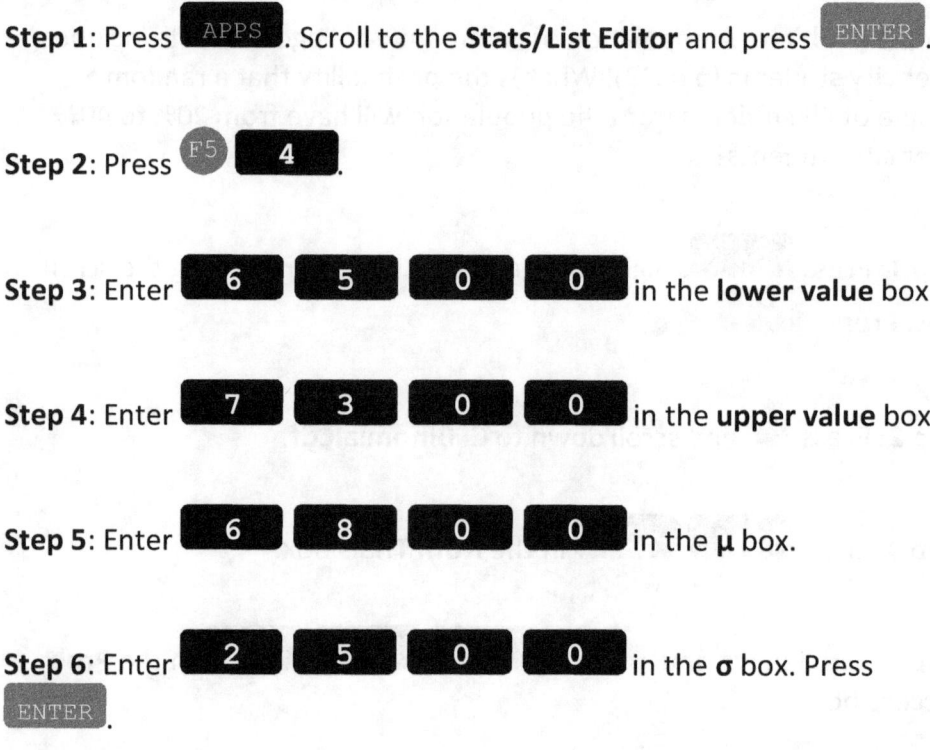

Step 1: Press APPS . Scroll to the **Stats/List Editor** and press ENTER .

Step 2: Press F5 4 .

Step 3: Enter 6 5 0 0 in the **lower value** box.

Step 4: Enter 7 3 0 0 in the **upper value** box.

Step 5: Enter 6 8 0 0 in the **μ** box.

Step 6: Enter 2 5 0 0 in the **σ** box. Press ENTER .

Step 7: Read the answer. **Cdf=.127018**. In other words, .013, or 13% of students earn between $6,500 and $7,300.

3. Finding Cut Off Points for a Top Percentage

Sample problem: Students at a certain college average 5 feet 8 inches (68 inches) tall. The heights are normally distributed, with a standard deviation of 2.5 inches. What is the value that separates the top 1% of heights from the rest of the population?

Step 1: Press **APPS** and use the scroll keys to highlight **Stats/List Editor**.

Step 2: Press **F5** **2** **1**. (This gets you to the Inverse Normal screen).

Step 3: Enter **.** **9** **9** in the **Area** box.

Step 4: Enter **6** **8** in the **μ** box.

Step 5: Enter **2** **.** **5** in the **σ** box.

Step 6: Press **ENTER**.

Step 7: Read the results: **Inverse=73.8159** means that the cut off height for the 99th percentile is **73.8159 inches**.

4. Draw a Normal Distribution Curve

Sample problem: Draw a normal distribution curve for student's salaries during a typical semester using a mean of $6,800 and standard deviation of $2,500. Shade the area that corresponds to salaries between $7,300 and $9,000.

Step 1: Press 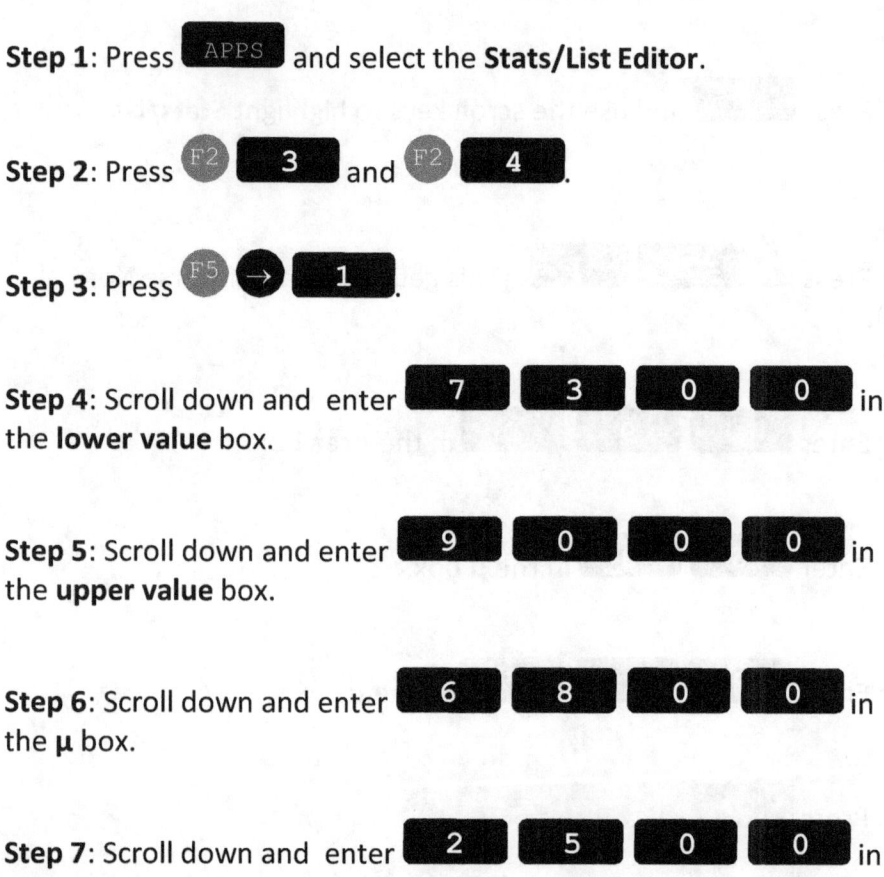 APPS and select the **Stats/List Editor**.

Step 2: Press F2 3 and F2 4.

Step 3: Press F5 → 1.

Step 4: Scroll down and enter 7 3 0 0 in the **lower value** box.

Step 5: Scroll down and enter 9 0 0 0 in the **upper value** box.

Step 6: Scroll down and enter 6 8 0 0 in the **μ** box.

Step 7: Scroll down and enter 2 5 0 0 in the **σ** box.

Step 8: Scroll down. Turn **Auto Scale** to "yes" by pressing the right scroll key, then the down scroll key to select yes. Press ENTER.

Tip: If you want to enter ∞ (infinity) as one of your lower or upper values, press the ♦ then CATALOG.

Z Scores

1. Find a Critical Value (Left-Tailed Tests)

Sample problem: find α = .012 for a left-tailed test on a standard normal distribution.

Step 1: Press **APPS**, scroll to the **Stats/List Editor**, and press **ENTER**.

Step 2: Press **F5** **2** **1**, to get to the **Inverse Normal** screen.

Step 3: Enter **.** **0** **1** **2** in the **Area** box.

Step 4: Enter **0** for the mean, μ, and **1** for the standard deviation, σ.

Step 5: Press **ENTER**.

Step 6: Read the result: the calculator should state "**Inverse = -2.25713**". This is your critical value.

Tip: If you are given a mean and standard deviation, enter them in place of 0 and 1 in Step 4.

Hypothesis Testing

1. Hypothesis Test on a Mean

Sample problem: Fifty years ago, middle school students at a local school had a mean height (μ) of 5ft 4 inches with a standard deviation (σ) of 2.4 inches. You want to test the hypothesis that the mean has increased over the last 50 years. You take a random sample of 30 students and find their mean is 65.11 with a standard deviation of 2.26569. At α=0.05, can you conclude that their mean height has increase based on this information?

Step 1: Press [APPS] and scroll to the **Stats/List Editor**. Press [ENTER].

Step 2: Write out the hypothesis. The null hypothesis is μ = 64. The alternate hypothesis (the one you are testing) is μ > 64.

Step 3: Press [2ND] , [F1] , [1] . Make sure the data entry method says "Stats." Press [ENTER].

Step 4: Enter the following values: μ_0 = 64, σ = 2.4, x = 65.1, n = 30. Scroll down to **Alternate Hyp** and choose **Hyp: μ > μ_o**.

Step 5: Press [ENTER] and read the results. The p value is **0.00603**. This is smaller than the significance level of 0.05, so you reject the null hypothesis and accept the alternate hypothesis: the mean has increased over the last 50 years.

Tip: If you get a domain error, clear your inputs from the **Stats/List Editor** by pressing [F1] , [8] .

2. Large Sample Hypothesis Testing (Proportion)

Sample problem: 62% of students agreed with the recent tuition increase. Test the claim that more than 62% of students would be in favor of the tuition increase today, at a significance level of 0.05. You have taken a simple random sample of 260 students, and 69.4% of them would be in favor.

Step 1: State the hypothesis and the alternate hypothesis. The hypothesis is stated in the question: 62% of students agreed with the tuition increase so H_o is p = 0.62. The alternate hypothesis (the one we want to test) is that more than 62% would vote for it: H_1 is p > 0.62.

Step 2: Make sure you can use the normal approximation to the binomial:

n × p = 260 × .694 = 180.**4**

180.4 > 10

and

n × (1 − p) = 98.8

98.8 > 10.

So we can use normal approximation.

Step 3: Press [APPS], scroll to the **Stats/List Editor** and press [ENTER].

Step 4: Press [2ND], [F1], [5]. This brings you to the 1-proportion z-test screen.

Step 5: Enter [0] [.] [6] [2] into the **p0** box.

Step 6: Press the scroll down key and 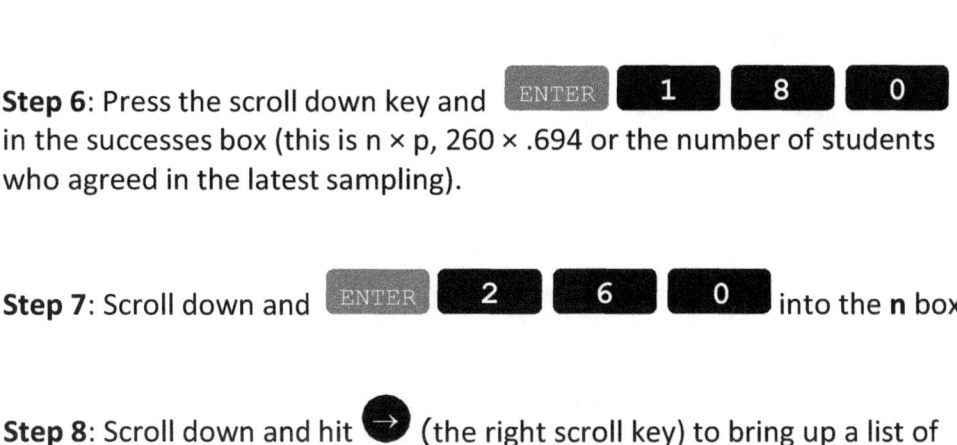 in the successes box (this is n × p, 260 × .694 or the number of students who agreed in the latest sampling).

Step 7: Scroll down and ENTER 2 6 0 into the **n** box.

Step 8: Scroll down and hit → (the right scroll key) to bring up a list of options. Scroll down to **prop>p₀**. Press ENTER .

Step 9: Scroll down to "Results." Use the scroll keys to select "Calculate." Press ENTER .

Step 10: Read the results: The P-Value given by the results screen is **0.008152**. This is smaller than your alpha level (0.05) so there is strong evidence that the new proportion will be larger than 63%, so you reject the null hypothesis and accept the alternate hypothesis.

Tip #1: If you get a domain error, you may need to clear the data in your list editor. Press ESC , F1 , 8 and try the steps again.

Tip #2: Choosing "Draw" instead of "Calculate" in Step 9 will give you a nifty graph of your result instead of a calculations screen. This is handy if you need to visualize something to learn it!

Warning: This calculating is for a one-tailed test (greater than). If you have a two-tailed test or a one-tailed test (less than) you'll need to change the option in Step 8 to **prop≠p** (two-tailed test) or **prop<p** (one-tailed test, less than).

Confidence Intervals

1. Confidence Interval for a Mean (Known Standard Deviation)

Sample problem: Fifty students at a Florida college have the following grade point averages: 94.8, 84.1, 83.2, 74.0, 75.1, 76.2, 79.1, 80.1, 92.1, 74.2, 64.2, 41.8, 57.2, 59.1, 65.0, 75.1, 79.2, 95.0, 99.8, 89.1, 59.2, 64.0, 75.1, 78.2, 95.0, 97.8, 89.1, 65.2, 41.9, 55.2. Find the 95% confidence interval for the population mean, given that σ = 2.27.

Step 1: Press `APPS` and scroll to **Stats/List Editor**. Press `ENTER`.

Step 2: Press `F1` `8`. This clears the list editor.

Step 3: Press `ALPHA` `)` `ALPHA` `9` `2` to name the list "CI2."

Step 4: Enter your data in a list. Follow each number with the `ENTER` key: 94.8, 84.1, 83.2, 74.0, 75.1, 76.2, 79.1, 80.1, 92.1, 74.2, 64.2, 41.8, 57.2, 59.1, 65.0, 75.1, 79.2, 95.0, 99.8, 89.1, 59.2, 64.0, 75.1, 78.2, 95.0, 97.8, 89.1, 65.2, 41.9, 55.2.

Step 5: Press `F4`, `1`.

Step 6: Enter "ci" in the "List" box: `ALPHA` `)` `ALPHA` `9`.

Step 7: Enter `1` in the **frequency** box. Press `ENTER`. This should give you the mean (xbar, the first in the list) = **75.033**.

Step 8: Press 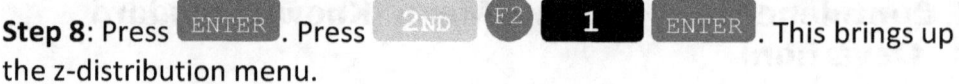. This brings up the z-distribution menu.

Step 9: Press 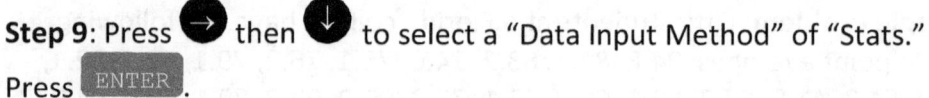 then ↓ to select a "Data Input Method" of "Stats." Press ENTER.

Step 10: Enter your σ from the question (in our case, 2.27), xbar from Step 7 (75.3033), n = 30 and the Confidence Interval from the question (in our example, it's .95).

Step 11: Press ENTER and read the results. The "C Int" is **{74.49,76.123}**. This means we are 95% confident that the population mean falls between 74.49 and 76.123.

2. Confidence Interval for a Mean (Unknown Standard Deviation)

Sample problem: A random sample of 30 students at a Florida college has the following grade point averages: 59.1, 65.0, 75.1, 79.2, 95.0, 99.8, 89.1, 65.2, 41.9, 55.2, 94.8, 84.1, 83.2, 74.0, 75.1, 76.2, 79.1, 80.1, 92.1, 74.2, 59.2, 64.0, 75.1, 78.2, 95.0, 97.8, 89.1, 64.2, 41.8, 57.2. What is the 90% confidence interval for the population mean?

Step 1: Press `APPS`. Scroll to the **Stats/List Editor** and press `ENTER`.

Step 2: Press `F1` `8` to clear the editor.

Step 3: Press `ALPHA` `)` `ALPHA` `9` to name the list "CI."

Step 4: Enter your data in a list. Follow each number with the `ENTER` key: 59.1, 65.0, 75.1, 79.2, 95.0, 99.8, 89.1, 65.2, 41.9, 55.2, 94.8, 84.1, 83.2, 74.0, 75.1, 76.2, 79.1, 80.1, 92.1, 74.2, 59.2, 64.0, 75.1, 78.2, 95.0, 97.8, 89.1, 64.2, 41.8, 57.2.

Step 5: Press `F4`, `1`.

Step 6: Enter "ci" in the **List** box: `ALPHA` `)` `ALPHA` `9`.

Step 7: Enter `1` in the **frequency** box. Press `ENTER`. This should give you the sample standard deviation, $s_x = 15.6259$, n = 30, and x (the sample mean) = 75.033.

Step 8: Press `ENTER`. Press `2ND` `F2` `2`.

Step 9: Press ⟶ then ⬇ to select a "Data Input Method" of "Stats." Press ENTER.

Step 10: Enter your x, s_x and n from Step 7. In our example, s_x = 15.6259. n = 30 and x = 75.033. Enter the Confidence Interval from the question (in our example, it's .9).

Step 11: Press ENTER and read the results. The **C Int** is **{70.19,79.88}** which means that we are 90% confident that the population mean falls between 70.19 and 79.88.

3. Confidence Intervals for a Proportion

Sample problem: In a simple random sample of 295 students, 59.4% of students agreed to a tuition increase to fund increased professor salaries. What is the 95% confidence interval for the proportion in the entire student body who would agree?

Step 1: Press [APPS] and scroll down to **Stats/List Editor**. Press [ENTER].

Step 2: Press [2ND] [F2] [5] for the **1-PropZInt** menu.

Step 3: Figure out your "successes." Out of 295 people, 59.4% said yes, so .694 × 295 = **175** people.

Step 4: Enter your answer from Step 3 into the **Successes,x** box. [1] [7] [5].

Step 5: Scroll down to **n**. Enter [2] [9] [5], the number in the sample.

Step 6: Scroll down to **C Level**. Enter the given confidence level. In our example, that's [.] [9] [5]. Press [ENTER], [ENTER].

Step 7: Read the result. The calculator returns the result **C Int {.5372, .6493}**. This means that you are 95% confident that between 54% and 65% of the student body agree with your decision.

Tip: If   u are asked for a folder when entering the Stats Editor, just press   It doesn't matter which folder you use.

Warning: Make sure your round your "success" entries to the nearest integer to avoid a domain error.

TScores

1. How to Use a T-Distribution

When you have a small sample size, or don't know the standard deviation of a population, you can't use the central limit theorem (which says that certain statistics like the sample mean will be normally distributed as long as you have a large enough sample size). In these cases, you need to use a T score, not a Z score. Unlike the Z scores, there isn't one table: there are many, and it depends upon the number of "degrees of freedom" (the number of independent observations) in your data set. For most T-distribution questions in elementary stats, you'll usually be given all of the information you need to plug into the calculator and retrieve the T score. You might be asked to find the area under a T curve, or (like Z scores), you might be given a certain area and asked to find the T score.

Sample problem: Find the area under a T curve with degrees of freedom 10 for P(1≤ X ≤ 2).

Step 1: Press APPS.

Step 2: Press ENTER ENTER.

Step 3: Press F5 for **F5Distr**.

Step 4: Choose 6 for **6:t Cdf**.

Step 5: Enter 1 in the box for **Lower Value**.

Step 6: Enter 2 in the box for **Upper Value**.

Step 7: Enter **1** **0** in the box for **Deg of Freedom, df**.

Step 8: Press `ENTER`. This returns the result **.133753**.

Sample problem: Find the T score with a value of 0.25 to the left and df of 10.

Step 1: Press `APPS`.

Step 2: Press `ENTER` `ENTER`.

Step 3: Press `F5` for **F5Distr**.

Step 4: Press **2** for **Inverse**.

Step 5: Press →.

Step 6: Press **2** for **Inverse t** and then press `ENTER`.

Step 7: Enter **0** **.** **2** **5** in the **Area** box.

Step 8: Enter **1** **0** in the **Deg of Freedom, df** box.

Step 9: Press `ENTER`. The calculator returns the result of **-.699812**.

Tip: For P(X ≥ a), enter 10^99 in the box for Upper Value and for P(X ≤ b), enter 10^99 in the box for Lower Value.

2. Find the Area Under a T Curve

Sample problem: Find the area under a T curve with degrees of freedom 10 for P(1≤ X ≤ 2).

Step 1: Press  .

Step 2: Press ENTER ENTER .

Step 3: Press F5 for F5 **Distr**.

Step 4: Choose 6 for **6:t Cdf**.

Step 5: Enter 1 in the box for **Lower Value**.

Step 6: Enter 2 in the box for **Upper Value**.

Step 7: Enter 1 0 in the box for **Deg of Freedom, df**.

Step 8: Press ENTER . This returns the result **.133753**.

Tip: For P(X ≥ a), enter 10^99 in the box for Upper Value and for P(X ≤ b), enter 10^99 in the box for Lower Value.

3. Find a T Score

Sample problem: find the T score with a value of .25 to the left and df of 10.

Step 1: Press `APPS`.

Step 2: Press `ENTER` `ENTER`.

Step 3: Press `F5` for **F5Distr**.

Step 4: Press `2` for **Inverse**.

Step 5: Press `→`.

Step 6: Press `2` for **Inverse t** and then press `ENTER`.

Step 7: Enter `0` `.` `2` `5` in the **Area** box.

Step 8: Enter `1` `0` in the **Deg of Freedom, df** box.

Step 9: Press `ENTER`. The calculator returns the result of **-.699812**.

Tip: For P(X ≥ a), enter 10^99 in the box for Upper Value and for P(X ≤ b), enter 10^99 in the box for Lower Value.

F Score

1. How to Find an F Distribution

There are two types of main problem you'll encounter with the F-Distribution: you might be asked to find the area under a F curve given numerator degrees of freedom (ndf), denominator degrees of freedom (ddf), and a certain range (for example, P($1 \leq X \leq 2$)), or you might be asked to find the F value with area to the left, a certain ndf and ddf (useful for finding critical values for hypotheses tests).

Sample problem: Find the area under a F curve with numerator degrees of freedom (ndf) 4 and denominator degrees of freedom (ddf) 10 for

P($1 \leq X \leq 2$):

Step 1: Press  .

Step 2: Press ENTER ENTER to get to the list entry screen.

Step 3: Press F5 for "F5-Distr."

Step 4: Scroll down to "A:F Cdf" and press ENTER .

Step 5: Enter 1 in the box for "Lower Value," then press ⬇.

Step 6: Enter 2 in the box for "Upper Value," then press ⬇.

Step 7: Enter 4 in the "Num df" box, then press ⬇.

Step 8: Enter [5] in the "Den df" box.

Step 9: Press [ENTER]. The calculator will return **.281** as the answer.

Sample problem: Find the F value with area to the left, with ndf = 5, ddf = 8, and an area of .99:

Step 1: Press [APPS].

Step 2: Press [ENTER] [ENTER] to get to the list entry screen.

Step 3: Press [F5] for "F5-Distr."

Step 4: Press [2] for "Inverse."

Step 5: Press [4] for "Inverse F...," then press [ENTER].

Step 6: Enter [.] [9] [9] in the "Area" box, then press [↓].

Step 7: Enter [5] in the "Num df," box, then press [↓].

Step 8: Enter [8] in the "Den df." box, then press [ENTER]. This returns the answer **63183**.

Tip: For P(X ≥ 1), enter 1 in the box for Lower Value and 10 ^ 99 in the box for Upper Value, and for P(X ≤ 1), enter 0 in the box for Lower Value, then enter 1 in the box for Upper Value.

Linear Regression

1. Linear Regression

Sample problem: Find a linear regression equation (of the form y = ax + b) for x-values of 1, 2, 3, 4, 5 and y-values of 3, 9, 27, 64, and 102.

Step 1: Press APPS then go to the **Data/Matrix Editor** screen using the arrow keys, then press ENTER.

Step 2: Scroll down to **3: New** using ⬇. Press ENTER.

Step 3: Press ⬇ to scroll to the **Variable** box. Enter your name by pressing the ALPHA key and using the alphanumeric keypad (the white letters to the top right of every key on the lower half of the calculator).

Step 4: Press ENTER ENTER.

Step 5: Fill your x-values in the first column (c1):

1	ENTER
2	ENTER
3	ENTER
4	ENTER
5	ENTER

Step 6: Scroll to the beginning of the second column by pressing ➡ and ⬆. Enter your y-values in the second column.

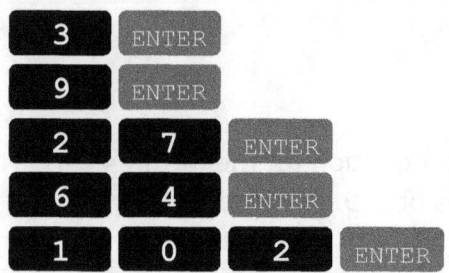

Step 7: Press F5. The calculator's cursor will flash on **Calculation Type**. Press → to bring up a scroll down menu. Use ↓ to select **5: LinReg**. Press ENTER.

Step 8: Press ↓ to the next row (x). Enter c1 using the alphanumeric keypad (ALPHA) 1). This tells the calculator where your x-variables are.

Step 9: Press ↓ to the next row (y). Enter ALPHA) 2 : this puts c2 into the space. Press ENTER. The calculator returns the value of a and b for you to enter into the equation y'=ax+b (in our equation, a = 25.3 and b = -34.9).

Step 10: Insert the calculator's result into the equation y' = ax+b: **y' = 25.3x + 34.9**

Two Populations

1. Confidence Intervals for Two Populations

Sample problem: A recent poll in a simple random sample of 986 women college students found that 699 agreed that textbooks were too expensive. Out of 921 men surveyed by the same manner, 750 thought that textbooks were too expensive. What is the 95% confidence interval for the difference in proportions between the two populations?

Step 1: Press  APPS, scroll to the **Stats/List Editor**, and press ENTER.

Step 2: Press  2ND F2 **6** to reach **2-PropZint**.

Step 3: Enter your values into the following boxes (Use "women" for population 1 (x1 and n1) and "men" for population 2 (x2 and n2)):

Successes, x1: 590

n1: 796

Successes, x2: 548

n2: 800

C Level: 0.95

Step 4: Press ENTER.

Step 5: Read the result. The confidence level is displayed at the top as **C Int { .0119,.10053}**. This means that your confidence interval is **between 1.19% and 10.05%**.

Tip: As long as you keep track of which population is x1/n1 and x2/n2, it doesn't matter which is entered in which box.

www.ingramcontent.com/pod-product-compliance
Lightning Source LLC
Chambersburg PA
CBHW071305170526
45165CB00003B/1423